Upstream Engineering and Flow Assurance Forum 2018

Core Programming Area at the 2018 AIChE Annual Meeting

Pittsburgh, Pennsylvania, USA
28 October - 2 November 2018

ISBN: 978-1-5108-7611-8

Printed from e-media with permission by:

Curran Associates, Inc.
57 Morehouse Lane
Red Hook, NY 12571

Printed by Curran Associates, Inc. (2019)

For permission requests, please contact AIChE
at the address below.

AIChE
120 Wall Street, FL 23
New York, NY 10005-4020

Phone: (800) 242-4363
Fax: (203) 775-5177

www.aiche.org

Additional copies of this publication are available from:

Curran Associates, Inc.
57 Morehouse Lane
Red Hook, NY 12571 USA
Phone: 845-758-0400
Fax: 845-758-2633
Email: curran@proceedings.com
Web: www.proceedings.com

TABLE OF CONTENTS

Some abstracts in this book refer to images that were not made available by the publisher

544713 Mitigation of Asphaltene Deposition by Dead Oil Recycle

Abstract:

Asphaltene deposition is a common flow assurance problem in the oil and gas industry. Asphaltene precipitation occurs when the pressure of a crude oil system falls below the asphaltene onset pressure (AOP) which may lead to subsequent deposition and plugging of the production tubing. The AOP curve is usually obtained experimentally and serves as an input to Equation of State (EoS) models. In this work, we highlight the importance of defining asphaltene precipitation by both the AOP curve and the amount of asphaltene precipitated. Additionally, we highlight the need for an effective mitigation technique to prevent asphaltene deposition. Thus, we propose a novel chemical-free mitigation technique, where dead oil is recycled to reduce asphaltene precipitation. The stabilization effect of the dead oil reinjection was verified using a thermodynamic model (PC-SAFT EoS), where a shift in the AOP curve is observed. Depending on the amount of reinjection and the crude oil characteristics, asphaltene precipitation can be eliminated by the reinjection of dead oil.

534595 Kinetics Study of Asphaltenes Adsorption Onto Hydrophilic Solid Surfaces

Abstract:

Large quantities of crude oil are left behind in some rock reservoirs after waterflooding in enhanced oil recovery processes. Asphaltenes, as the heaviest and most polar indigenous component of crude oils, is believed to have a profound influence on wettability changes of rock reservoir. The physics of how asphaltenes adsorb onto originally water-wet surfaces and alter their wettability is still poorly understood. Quantifying the effect of asphaltenes adsorption on wetting and the interaction between asphaltenes and rock surfaces remains a big challenge.

This study investigates the controlling asphaltenes adsorption mechanisms onto hydrophilic surfaces using the experimental results from quartz crystal microbalance with dissipation (QCM- D), and examines the wetting property changes caused by the adsorbed asphaltenes layer. Within the most common types of hydrophilic rock surfaces, calcium carbonate and silicon dioxide are tested.

The maximum surface coverages obtained from QCM-D experiment at different concentrations indicate a two-step process, which can be successfully interpreted with a multilayer adsorption model, i.e. the so-called Brunauer, Emmet & Teller (BET) model. Both the equation of state and adsorption isotherm for multilayer adsorption can be derived from statistical mechanics theories. Using an optimization method that was developed, it was found that the optimized BET adsorption isotherm curve not only fits the experimental data well at low concentrations like the Langmuir adsorption isotherm does, but also approaches to the experimental data at high coverage where the Langmuir isotherm diverges. The water wettability of the crystal sensors are examined before and after asphaltenes adsorption to investigate the effect of the adsorbed asphaltenes layer, which is found to increase the water to crystal contact angle.

534056 The Effect of Cyclic Molecules on the Gelation Characteristics of Polydisperse n-Alkane Systems

Abstract:

Wax deposition in the pipelines is a very complex and costly problem present in the petroleum industry. In many cases, the major components of the wax are n-alkanes. However, other components in the crude oil can also influence the gelation properties. Although a significant amount of research has been done on the effects of asphaltenes, resins and n-alkanes on crude oil gelation, the role of aromatics in wax gelation has been less explored. To gain more fundamental insights on n-alkane gelation, the effects of cyclic molecules on n-alkane gelation were studied. Model waxes composed of one or two alkanes selected from three n-alkanes C28, C32 and C36, were mixed with either a single-ringed aromatic compound durene and polycyclic compound phenanthrene in a dodecane (C12) solvent. Rheometry experiments using oscillatory stress and specified cooling rate was used to assess the gel points of each model oil sample. Based on the gel point results, it was found that solubility, molecular size and amount each species in the model oil all influence the gelation, both individually and together. The least soluble species that crystallized first usually determined the gelation in terms of gel points. However, the molecular size difference can have an influence on the dispersant effect that can hinder gelation for polydisperse n-alkane systems. It is hypothesized that the aromaticity plays an important role in the gelation process by fostering the self-aggregation of shorter-chained n-alkanes, preventing them to interact with longer-chained alkanes.

536115 Removal of Asphaltene from Crude Oil through Electrodeposition Method

Abstract:

The deposition of asphaltene on interior pipe surfaces is considered a challenging flow assurance issue. To prevent this, chemical dispersants are often added, which increases transportation costs and introduces environmental concerns. By exploiting the net charge of the asphaltene molecule, they can be effectively removed from the flow stream by means of electrodeposition before transportation to prevent later deposition. To evaluate this concept, electrodeposition of asphaltene from a synthetic oil using varying electric field strength and solvent type were studied. Both native charge (positive) and induced charge (negative) of asphaltene were observed in this study, and the inducement of charge was realized with higher current density. Asphaltene solubility in the synthetic oil was increased by increasing the polar/non-polar solvent ratio revealing a higher conductivity of the synthetic oil. As a result, the synthetic oil with higher polar solvent content provided higher current density resulting in deposition onto the anode, inferring an induced negative charge of the asphaltene. In addition to electric field strength and synthetic oil composition, the asphaltene concentration and synthetic oil viscosity affected the electro-deposition. The process was also employed in multiple cycles with continuous asphaltene deposition; the recovery rate was over 90 wt.% in the first three cycles and the deposition gradually decreased as the deposited asphaltene appeared to diminish the electric field strength.

526245 The Influence of the Reservoir Acidity on Asphaltenes Dissolution in Aromatic Solvent Using Microsystems with *in Situ* Spectroscopy

Abstract:

Asphaltenes give rise to flow assurance issues in upstream petroleum and natural gas production.[1,2] The molecular structure of asphaltene includes an aromatic core, heteroatoms (*e.g.*, N, O, and S), and a trace amount of metals (*e.g.*, V, Ni). Their complex structures lead to five different types of molecular interactions: van der Waals, acid-base, metal coordination, hydrogen bonding, and p-p stacking.[3] The fact that the structure and composition of asphaltenes vary from reservoir-to-reservoir adds to the complexity of flow assurance problems. Molecular-level information on asphaltenes that explains production-scale phenomena is critical to their remediation, and microfluidics with *in situ* spectroscopy are appropriate experimental methods.

In the present work, the deposition and the dissolution mechanisms of asphaltenes in aromatic solvent were studied at ambient temperature in a micro-packed-bed reactor with the help of in-line UV-Vis spectroscopy and *in situ* Raman spectroscopy. Measurement of the bed occupancy yielded the asphaltene distribution within the porous media. Zeolite (H-ZSM5) with different ratio of Al_2O_3/SiO_2 were also injected to study the influence of surface acidity of porous media. According to our results, the dissolution percentage of asphaltene in xylenes increases from 20.15 wt% to 51.00 wt% as the Al_2O_3 content increased. Measurements of the asphaltenes sheet-size distributions provided insight on the interaction mechanisms. Asphaltene molecules with larger sheet sizes tend to have stronger affinity with aromatic solvents than those with smaller sheet sizes.[4] The mean sheet size was observed to decrease from 2.97 ± 0.25 nm to 2.74 ± 0.26 nm. Our findings support that p-p stacking is significant among the different interactions of asphaltenes in the aromatic solvent.[5] General guidelines based on our observations could be prescribed for the design of asphaltenes remediations in the near-wellbore region.

References:

1 K. Akbarzadeh, A. Hammami, A. Kharrat, D. Zhang, S. Allenson, J. Creek, S. Kabir, J. Jamaluddin, A. Marshall, R. Rodgers, O. Mullins and T. Solbakken, *Oilf. Rev.*, 2007, 22–43.

2 O. C. Mullins, *Annu. Rev. Anal. Chem.*, 2011, **4**, 393–418.

3 M. R. Gray, R. R. Tykwinski, J. M. Stryker and X. Tan, *Energy and Fuels*, 2011, **25**, 3125–3134.

4 E. Rogel, *Energy & Fuels*, 2000, **14**, 566–574.

5 J. J. Adams, *Energy and Fuels*, 2014, **28**, 2831–2856.

536661 Effects of the Presence of Water in Cold Restart of Waxy Oils

Abstract:

Recent work has shown the validity of cold flow, a technology to transport waxy crude oils below their wax appearance temperatures (WAT). The thermal flux that results due to the difference between the temperature of the transported oil and ambient is minimized in the implementation of cold flow – thus significantly lowering deposition. Restarting pipelines carrying waxy oil under cold flow is an important consideration. Previous work from this group had shown that restart pressures required for pipelines shutdown under cold flow at a given temperature were lower than pressures required to restart conventionally transported oils (above WAT). In this paper we examined the applicability of this conclusion when large amounts of water were present with the oil.

A model oil consisting of 7% wax and about 0.5 wt% of an appropriate surfactant was used in experiments with water. Maximum amount of water in emulsions was 40%. A reproducible rigorous mixing protocol was developed to create emulsions stable over the course of the experiments. Comprehensive rheological measurements were performed on water-free and emulsified samples using cone and plate geometry. The WAT and gel point were measured using the oscillation method with a slow cooling rate. We used creep methods to measure the static yield stress for gels at 5 °C formed under hot/cold flows. Yield stress measured for emulsion gels is largely lower than that for water-free gels. We studied the effects of roughness of the cone and plate surface and found that higher yield stress was observed using the roughened geometry.

Restart studies were conducted in a bench-scale, 0.4-inch diameter, 4-foot long flow loop. The distinguishing feature of the flow loop was a conditioning loop consisting of a scraped heat exchanger to prepare the cold slurry for restart experiments. Water-free cold flow experiments were first performed to establish a baseline. The measured restart pressures for cold flow restarts at 5 °C when shutdown temperatures were 5-10 °C below the WAT were approximately three times lower than when "hot" restart (cooling from above the WAT to the designated temperature) were used. The restart pressures decreased as progressively lower temperatures were used at shutdown. Restart with emulsions showed that even lower restart pressures were needed for both cold and hot restart. The flow loop experiments results agreed with the rheological measurements that emulsion gel is weakened comparing with the water-free waxy gel. The dependence of lower pressure for progressively colder restarts observed for water-free systems was minimal to non-existent and consistently lower restart pressures were observed. Observations of gelled emulsions under the microscope at different temperatures confirmed the disorganized structure of the gel with water present. We believe this morphology of the gel was primarily the cause of the lower restart pressures observed for emulsified waxy oils.

515234 Avoiding Flowline Plugging: Hydrates Pressure Drop Signature, Plugging Onset Prediction and Deposition Mechanism

Abstract:

Hydrate formation, if not properly monitored and controlled, may lead to pipeline blockage. In order to predict pipeline blockage onset and to develop effective hydrate plugging mitigation strategies, flow assurance engineers must understand the indicators and precursors of hydrate plugging onset and the corresponding system response. Hydrate transportability and plugging onset strongly depend on carrier fluid properties and flow geometry.

This paper presents work done towards understanding indicators of hydrate plugging onset and the governing plugging mechanisms. Extensive multiphase flowloop studies were conducted to establish the pressure drop signature, temperature response, and visual observation of plugging behavior of the hydrate transporting systems. It was observed that for low viscosity carrier fluids, i.e., low viscosity oils, hydrate transportability, plugging onset and plugging mechanisms were strongly dependent on multiphase flow effects. Specifically, the main factors included liquid loading, watercut, mixture velocity and presence or absence of additive. General observation for these low viscosity systems included plugging risk increasing with: increasing watercut, decreasing liquid loading, absence of additives, low mixture velocity and increased sub-cool. Other factors that increased the risk included salinity, pressure and rate of cooling. Onset of plugging was deduced from pressure drop increasing or plateauing or increasing with decreasing mixture velocity. On the other hand, onset of bedding led to scattered temperature readings. Overall, hydrate deposition and plugging result from severe agglomeration leading to particle-wall interaction.

The second part of the paper presents results for the high viscosity carrier fluids, i.e., high viscosity oils. These systems were characterized by very high pressure losses. These systems are very sensitive to temperature change, i.e., cooling into hydrate zone was accompanied by huge pressure drop increase. Visual observations through viewport showed very little activity close to the pipe wall. It was hypothesized that flow in these systems was governed by significant boundary layer effects, i.e., as viscosity further increased due cooling, the laminar boundary transformed into creeping flow and finally stalled. In these systems, hydrates transportability and bedding strongly depended on emulsion properties. These systems were characterized by significant shear heating and difficult restarts.

Overall, hydrate deposition and plugging result from formation of strong emulsions with increasing yield stresses that lead stalled flow. Results showed that use of anti-agglomerants improves transportability of hydrates.

Keywords: rheology, particulate flow, slurry, flow regime, flow assurance, emulsions

Solids Flow, Handling, and Processing

523365 Existence of Supersaturation and Its Effect on Wax Deposition Behavior

Abstract:

Accurate prediction of wax deposition behavior in oil and gas production is necessary to design optimal flow assurance strategy. Current mechanistic models used for predicting wax deposition phenomenon have high uncertainty especially under turbulent flow conditions. This high uncertainty in prediction may be attributed to the incorrect assumptions of the underlying mechanisms that may not represent the process. Microscopic observations of flowing crystals above the interface validates the existence of partial supersaturation in the boundary layer and interface. A new model is proposed with the incorporation of the effect of supersaturation as a correction factor for the existing wax deposition model. This model is validated with flow loop data at various conditions.

540773 2018 Outlook for Energy: A View to 2040

Abstract:

The Outlook for Energy is ExxonMobil's global view of energy demand and supply through 2040. It also highlights the dual challenge of providing affordable energy to support prosperity while reducing environmental impacts – including the risks of climate change. We use the Outlook to help inform our long-term business strategies and investment plans.
The Outlook anticipates significant changes through 2040 across the world to boost living standards, reshape the use of energy, broaden access to abundant energy supplies, and accelerate decarbonization of the world's energy system to address the risks of climate change.

- Global energy needs will rise about 25 percent over the period to 2040
- Electricity demand will nearly double in non-OECD countries
- Worldwide electricity from solar and wind will increase about 400 percent
- Natural gas will expand its role, led by growth for electricity generation
- Growth in oil demand will be driven by commercial transportation and chemical needs
- Efficiency gains and growing use of less-carbon intensive energy sources will contribute to a nearly 45 percent decline in the carbon intensity of global GDP

Energy is fundamental to modern life, and as the world's population approaches 9 billion people in 2040, we are challenged to improve living standards everywhere. Understanding the factors that drive the world's energy needs – and likely choices to meet those needs – is the mission of the Outlook. By sharing the Outlook with the public, we hope to broaden that understanding. Energy matters to everyone, and we all play a role in shaping its future.

541543 Energy Decarbonisation Scenarios

Abstract:

With the energy sector being associated with 2/3 of the annual greenhouse gas emissions, decarbonizing the energy is a key step to reduce their level. The National Determined Contributions (NDC's) announced so far by the participating countries do not meet the objectives of the Paris-climate agreement, as far as reaching a peak in emissions in the near future or decreasing to a level that is compatible with the stated long-term targets.

Using different scenarios for energy demand and supply, options for energy sector decarbonisation are evaluated. Meeting the 2-degree target requires a major transformation in the power sector, efficiency improvements – specially in the buildings and in the industry, and a revolution in transport towards fuel and modal shifts. With a significant increase in renewable energy, system integration and optimization will also be key in reducing the cost of the energy transition. Meeting more ambitious climate goals ("well below 2 degrees") will push further the technological innovation and implementation requirements throughout the entire energy value chain. A scenario aiming at zero net-emissions by 2060 is presented, along with a comparison to the 2-degree case.

542625 Fundamental Research Needs to Advance Energy Technologies

Abstract:

The U. S. Department of Energy (DOE), Office of Science, Office of Basic Energy Sciences (BES) supports fundamental research in chemical and materials sciences to provide the foundations for new energy technologies and to support DOE missions in energy, environment, and national security. This presentation will discuss opportunities for fundamental research to impact DOE's energy mission "to catalyze the timely, material, and efficient transformation of the nation's energy system and secure U.S. leadership in energy technologies" with a focus on the way we generate, store and use energy nationally. I will provide an overview of BES strategic planning over the past decade that identified priority research directions for advancing energy applications, highlight key scientific advances in these areas, and discuss some future opportunities for modern science, particularly chemical sciences, to accelerate the transformation of the U. S. energy portfolio.

540877 The Current State of the Knowledge: Wax Deposition Modeling and Up Scaling Challenges

Abstract:

Wax deposition inside the transportation pipelines is one of the major flow assurance issues for offshore production systems. This deposit may not only get thicker, but also get harder over time. As it gets harder, the deposit removal process becomes more difficult. Proper understanding of wax deposition behavior is required to develop accurate deposition models and cost effective management strategies. Available models rarely achieve acceptable agreement with experimental data without fitting parameters, to the extent that more than 100% uncertainty is common. It is necessary to identify the sources of uncertainty to improve the model capability and up scaling efforts. This presentation will focus on the severity of wax deposition and current state of the knowledge and limitations on wax modeling in upstream pipelines.

534522 The Influence of Surfactants and Nanoparticles on Hydrate Formation in Water-in-Oil Emulsions

Abstract:

Clathrate hydrates are non-stoichiometric, ice-like crystalline solids that can lead to plugging of crude oil pipelines. Crude oil is a complex hydrocarbon mixture that includes asphaltenes, aromatics, naphthenes, resins, and paraffins. Although, the effect of nanoparticles and surfactants on hydrate formation has been investigated, the synergistic effects of surfactant and nanoparticles on hydrate formation are not well understood. Elucidating the synergistic effects of surfactants and nanoparticles on hydrate formation would improve the hydrate prediction strategies.

A fundamental investigation using model oil was carried out to determine the effect of interactions between silica nanoparticles and surfactants on hydrate formation in water-in-oil emulsions. Cyclopentane was used as the hydrate-forming component. A DHR-3 stress controlled rheometer was used to examine the rheological behavior of hydrate forming emulsions. In order to elucidate the importance of rheometer geometries, rheological behavior of hydrate slurries using standard (concentric cylinder) and non-standard (vane and helical ribbon) geometries was quantified. An Olympus BX53 polarized optical microscope was used to quantify droplet size distribution.

In addition to the wettability of nanoparticles, the mode of addition of nanoparticles played a critical role in determining the droplet size of emulsions, flow behavior of emulsions and hydrate induction times. Partially hydrophobic nanoparticles enhanced hydrate formation as compared to highly hydrophobic nanoparticles upon the addition of nanoparticles to surfactant containing oil phase before emulsion formation (pre-mixing). On the contrary, partially hydrophobic nanoparticles delayed hydrate formation as compared to highly hydrophobic nanoparticles upon the addition of nanoparticles to surfactant stabilized water-in-oil emulsions (post-mixing). Confocal microscopy was used to determine the spatial distribution of nanoparticles. Furthermore, interfacial tension experiments were performed in order to understand the surfactant-nanoparticles interactions at the oil-water interface.

534907 Prospects of Amino Acids and Ionic Liquids As a Potential Gas Hydrate Inhibitors for Offshore Flow Assurance

Abstract:

The gas hydrates are crystalline ice-like compounds that are formed under high pressure and low-temperature conditions in the offshore subsea lines. The formation of gas hydrates in the subsea lines is likely to cause unexpected operations shut-down, maintenance issues, safety issues and pipeline rapture in the worst case scenario. Therefore, the formation of hydrates is considered a major flow assurance issue in the oil and gas industry.

Annually, it is reported that the oil and gas industry spends over 200 million US $ on the purchase of chemical inhibitors like methanol and mono-ethylene glycol, to prevent the risk of hydrate formation by shifting the hydrate vapor liquid equilibrium (HVLE) curve to lower temperatures. However, these thermodynamic inhibitors are required in high dosage (> 30 wt%), they are toxic, flammable and there are environmental constraints associated with the disposal of these inhibitors. Therefore, the industry is in the quest of finding hydrate inhibitors that are environmentally benign, economical, required in low dosage and easily disposable.

Amino acids are biological compounds that naturally exist in nature and are considered as building blocks of life. The amino acids due to their inherited eco-friendliness and enhanced biodegradability do have a potential to act as the hydrate inhibitors. Ionic liquids are organic salts that have a low melting point and negligible vapor pressure. Ionic liquids are widely being used in green chemical processes, as many classes of ionic liquids are considered to be environmentally benign. Therefore, in this work, the prospect of amino acids and ionic liquids as future potential hydrate inhibitors have been experimentally tested at different pressures and concentrations.

All the experimental work has been conducted using pure methane gas and rocking cell assembly (RC-5) under diverse operating conditions (40-120 bars). The group has tested five commercially available amino acids and four specially synthesize ammonium based ionic liquids as hydrate inhibitors on pure methane gas. The dual functional behavior of ionic liquids and amino acids has also been investigated and the results obtained have been compared with the conventional hydrate inhibitors like methanol and mono-ethylene glycol. Accordingly, the synergistic effect of a water soluble polymer on the hydrate inhibition effectiveness of amino acids and ionic liquids has also been tested. This work provides valuable insight for the industry flow assurance experts and academic researchers.

Acknowledgement

This work was made possible by GSRA # 2-1-0603-14012 from the Qatar National Research Fund (a member of Qatar Foundation). The statements made herein are solely the responsibility of the authors.

515236 Flow Assurance Studies with Multiphase Flowloop: High Viscosity Carrier Fluids and Viscous Heating in Hydrate Transporting Systems

Abstract:

Over the life of a production well the GOR declines as the watercut increases. These changes result in high viscosity dispersion and even emulsions. As the flow assurance discipline shifts from preventing to managing hydrate formation, it is imperative for the flow assurance to fully evaluate the feasibility of hydrate transportability and the associated risks. Over the past four decades, tremendous work has been done towards understanding the impact of hydrate formation on flow properties. However, most of this work has been focused on low viscosity oils were flow assurance engineers used additives to improve phase dispersion and minimize agglomeration. Little efforts have been directed to high viscosity oils. These oils tend to form stable dispersion which poses two main challenges: high viscosity which may stall flow; high shear heating resulting from viscous friction and poor heat dissipation. Shear heating plays an important role in fluid flow because of the coupling between the energy and momentum equations. The generated heat leads to local temperature increase especially in the vicinity of pipe wall. This may have several consequences such as: local decrease in viscosity which may dramatically change the temperature and velocity profiles; affect precipitation and deposition of wax; affect solids concentration gradient and plugging mechanism for the case of hydrates. To this end, flow assurance engineers are pursuing use of drag reducing agent to minimize the effects of high viscosity fluids.

This paper presents preliminary work done towards understanding the shear heating phenomenon in oil transporting system. In addition to very high pressure losses, the systems are characterized by high shear heating effects. Four fluids were used to investigate this behavior and results showed that the temperature difference between the flowing fluid and the coolant in the annulus increased with fluid viscosity and shear rates.

The second part of the paper presents hydrate slurry flow with high viscosity fluids. Results showed that pressure drop and shear heating significantly increased during cooling into hydrate region. Shear heating continued to increase with increasing hydrate volume fraction (HVF). At a certain HVF, pressure losses hit a maximum and started decreasing. Shear heat was responsible for the change in pressure drop profile. This suggests that shear heating affects solid transport and deposition mechanism.

Lastly, the paper will present preliminary work towards use of additives to suppress the effects of high viscosity on oil transporting systems.

Iron Sulfide Scale Removal from Production Wells By New Chemical Formulation: A Field Application in a Sandstone Reservoir in Egypt

Eng. Emad Hamdy, Eng. Mohamed A. Bakr, Sapesco Co.

Abstract

Iron sulfide scale is a significant problem in the oil and gas industry whereas the iron sulfide deposition has an adverse impact to the production operations. The scale is present in various species with different iron to sulfur ratios. Iron sulfide species with high iron content is soluble in acids, whereas those rich in sulfur are almost insoluble in acids.

East ZEIT oil wells have been produced for 20 years in Egypt. Scales with iron sulfides as the major components form in nearly every producing well. These scales have minor compositional but form at differing rates and thicknesses in the lower half of tubing string. Well East ZEIT-13 was completed initially in Nubia formation, which is communicated with another formation called ''Lower Senonian', whereas the drive mechanism is active water-drive.

The formation is sandstone with the following specs: Lose, pale yellow, quartz, very fine to fine-grained, sub-rounded to rounded, and moderately sorted sand grains. The reservoir temperature is 300°F. The well was producing using a jet pump (JP), the production stopped due to scale formation. A workover started to replace the completion. The JP was used again to produce the well with very low productivity; +/-0.06 bbl/psi. The power fluid of the JP is oil with 1% water cut. The scale covered all the casing area below the tubing end until PBTD (Plug Back Total Depth).

A scale sample was obtained and analysis of the scale through XRD showed that the scale composition is as follows; 36% Marcasite (FeS_2), 21% Greigite (Fe_3S_4), 17% Pyrite (FeS_2) and 26% Siderite ($FeCO_3$). Both pyrite and marcasite have a very low solubility in HCl, and the only available method of removal in the oil industry is milling. The milling operation failed to solve the problem, therefore, the production was stopped until finding an alternative solution.

An effective treatment strategy proposed to remove the scale efficiently; the technical proposal was designed taking into consideration the complicated nature of the scale. Furthermore, the proposed treatments were conducted and placed using coiled tubing (CT).

The first application using the new chemical solution has been successfully carried out and the job results were outstanding. Production has been returned with an increment of 50% above its original rate. The well is producing natural flow without using the JP, and efficient cost /Bbl.

Introduction

Scale is secondary precipitation of mainly inorganic chemical compounds due to fluid flow systems partially man-made (Bajammal et al. 2013). It deposits due to incompatibility between brine present in the reservoir and production flow system. Scale precipitates as a result of the physicochemical properties changes such as pressure, temperature, pH, and salinity of the brine (Mackay 2003; Gomes et al. 2012; Quintero et al. 2018).

The scale can deposit inside the reservoir formation, the wellbore, and in the surface equipment. Scale deposition will restrict fluid flow, limit oil and gas production, and cause damage to downhole equipment's (Mackay 2003; Farrokhrouz and Asef 2010).

Iron sulfide can be deposited in different forms. In sour wells, scale analysis shows the presence of several types iron sulfide scales such as; Mackinawite, troilite, pyrrhotite, pyrite, and marcasite (Ford et al. 1992). The type of iron sulfide scale depends on different parameters, including temperature, water salinity, and the presence of other gases.

Producers or injectors that has sulfide-containing fluids will precipitate iron sulfide scales. Iron sulfide scale is present in sour wells, and seawater injectors that have sulfate-reducing bacteria (SRB). The low H_2S concentrations wells doesn't mean less scale comparing to high H_2S concentrations wells.

The iron source determines the amount of iron sulfide scale. The iron source can be corrosion process, formation water, and formation mineralogy and the precipitation of iron sulfide is more dependent on the pH and salinity of the water than on H_2S concentrations (Ford et al. 1992; Leal Jauregui et al. 2007).

To effectively design an acidizing treatment, scale compositions and types should be defined. Iron sulfide has different types, and each one has its own specific solubility. Several methods that have been successfully used to control iron precipitation in sweet wells but it don't work in the presence of hydrogen sulfide. Chelating agents, and reducing agents fail to prevent precipitation of iron with H_2S to form iron sulfide and, in some cases elemental sulfur. Hydrochloric acid can be used to dissolve iron sulfide deposits with sulfur/iron ratio of one or less (Nasr-El-Din et al. 2000; Nasr-El-Din and Al-Humaidan 2001; Yap et al. 2010). HCl reaction with iron sulfide produces hydrogen sulfide.

$$FeS + 2HCl \rightarrow H_2S + FeCl_2 \quad (1)$$

Iron sulfide scale dissolving is a difficult process for several reasons. Iron sulfide is an oil-wet material so depending on the well type, iron sulfide scale can be covered with oil, condensate, or biomass. These materials will inhibit acid reaction with the scale. Another issue is the continuous exposure of FeS to H_2S will precipitate iron sulfide scale with sulfur/iron ratio close to 2. These scale types are not soluble in different acids (Ford et al. 1992). A third issue is the reaction produces H_2S as shown in **Eq. 1** which is a toxic and corrosive gas, and should be removed from the system as soon as possible.

The objective of this paper is to present the application of a new iron sulfide scale dissolver on actual field case on sandstone reservoir in Egypt. Laboratory measurements were first conducted to examine the solubility of the scale using the dissolving agent. An oil well was then treated with the chemical and the production enhancement was monitored with time.

Background

Nubia is a good quality massive sandstone reservoir with a low volume of shale contents and is supported by strong water drive mechanism. Production from Nubia started in April 1986 with 2,500 bbl/day from the first well and reached to the maximum peak of production rate about 17,200 bbl/day from nine wells in October 1994. As a result of high water cut, the oil flow rate decreased to 700 bbl/day from two wells in late 2012 (Gamal et al. 2013).

Nubia sandstone formation is active water drive reservoir. The Nubia has been subdivided into four major zones from youngest to oldest: the transitional Nubia (TZ), MN, M1 and M2 zones. The age of the units ranges from Paleozoic to mid-Cretaceous.

Well A-13 is a vertical well ceased flowing naturally due to high water cut and Jet Pump (JP) was used for more than 13 years. From time to time; the well was suffering from scale accumulation problems, which extremely aggravated with the JP work. Workover was conducted to retrieve the completion with the JP. Scales were found on the casing; which restricted the workover operations. Running new completion equipped with sliding sleeves. JP production trial was 120 bbl/day with 80% water cut. Hard scales were found accumulating starting from the end of the tubing and all perfs were plugged by a hard scale. Productivity index decreased to 0.06 bbl/day/psi. Well, productivity index was measured by a downhole gauge during JP Inaccessible Wellbore. **Fig. 1** shows schematically for well A-13 completion.

CURRENT COMPLETION & WELL INFORMATION	
Original Elevations:	S4.5' RKB – THF

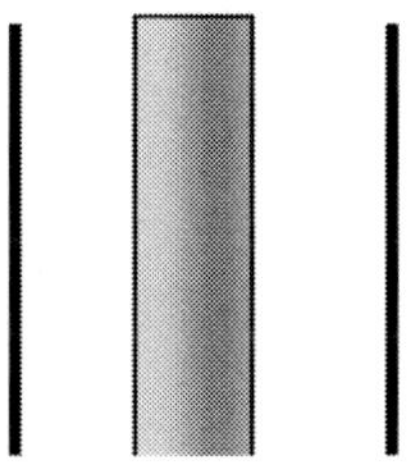

	116' RKB – MSL
Water Depth:	241 ' MSL
Target Formation:	Nubia 1 L. Senonian
Well PBTD:	11,383 ft MD RKB
TOF:	11,068 ft MD RKB (WLD in Jan 2017)
Production Casing:	7" , 32ppf, L-80, BTC@7,856' (10=6.094", Drift=5.696")
	7", 49.5ppf, C-90, HYD CST-4@7,856'-11 ,792' (ID=5.54",
	Drift=5.4IS")
	7", 29ppf, NK-95T, BTC@II ,792'-1187S' (ID=6.184",
	Drift=6.0S9"),
	Estimated T.O.C. +1- 6,000'
Tubing Head Spool:	FMC TCM ETII"' Sk x 71116" SM
Tubing Hanger:	FMC TC I A EN 71116" x 3Y," FF Material Class 3y," New Yam
Christmas tree:	3'1." 5000 psi
Tree Connections (Wing Valves):	3 Yo" flange, SK
Top Tree Connection:	Y I," EUE Internal Lift Thread
Current Production Mode:	Jet Pump
Current Packer:	3 1/2" X 7" Yam Top PHL Hydraulic PKR (29-35) @ 6,083' RKB
Fluid:	OSG=0.86, WSG= I.IS, 80% WC
Reservoir Pressure:	54S0 psi (9.1 ppg equiv. at 11500' TVDss)
Reservoir Temperature:	300 F
Current Tubing String:	3 1/2" Tubing, 9.2 ppf, L-80, JFE Bear
Burst:	10,160 psi, Yield Strength: 207 Klbs,
Collapse:	10,540 psi, 10=2.992"", Drift=2.867"
SSD Status:	Closed

Fig. 1 - Well A-13 completion after the workover.

Water samples were taken from the well production and analyses. Table 1 gives the analysis of the formation water.

Date	Cations Concentrations, ppm							Anions Concentrations, ppm			TDS, ppm
	Sodium	Calcium	Potassium	Magnesium	Iron	Barium	Strontium	Chlorides	Sulfates	Bicarbonate	
Mar-95	69,165	30,100		4,620	1,200			174,783	70	80	280,01
Jul-95	81,094	36,793	5,700	1,112	618	100	1,174	200,043	134	70	327,53
Apr-97	81,200	46,800		1,186	1,354	154	1,145	230,750	276	118	362,98
Dec-97	63,024	48,536		748	750			183,758	225		298,24

Mar -98	78,372 1	48,377		316	675			208,882	17	146	337,13

Table 1- Analysis of the formation water from well A-13.

Laboratory Measurements

To identify the scale composition, a scale sample was taken from downhole at 11,000 ft. and an XRD analysis was conducted to identify the scale minerals. Table 2 shows the mineral analysis for the scale samples.

Mineral	Marcasite, FeS_2	Greigite, Fe_3S_4	Pyrite, FeS_2	Siderite, $FeCO_3$
Concentration, wt%	36.09	20.6	17.09	26.22

Table 2- XRD analysis of the scale sample.

Solubility tests were conducted at room temperature (77°F) and at downhole conditions of 300 °F and 1000 psi for scale sample collected from a depth of 11,000 ft (**Fig. 2**). At room conditions, 5.5 g of the scale was soaked on 125 cm^3 of the scale removal on a glass jar for 16 hrs. The solubility was calculated as a function of the weight loss to be 73 %. At downhole conditions, High-pressure and the high-temperature visual cell were used on the test. 2.1 g of the scale was loaded on the HPHT cell and 50 cm^3 of the scale removal was then added. The temperature was set at 300 °F and the cell pressurized using nitrogen to 1000 psi. Table 3 shows the solubility test results of 73 and 77% of the initial scale weight at room and downhole conditions, respectively.

Initial scale weight, g	**Scale Removal Volume, cm³**	**Soaking Time, hr**	**Temperature, °F**	**Pressure, psi**	**Final Weight, g**	**Solubility, wt%**
5.542	125	16	77	14.7	1.5135	73
2.0941	50	16	300	1000	0.4857	77

Table 3- Solubility results of the scale samples with the new scale dissolver.

A

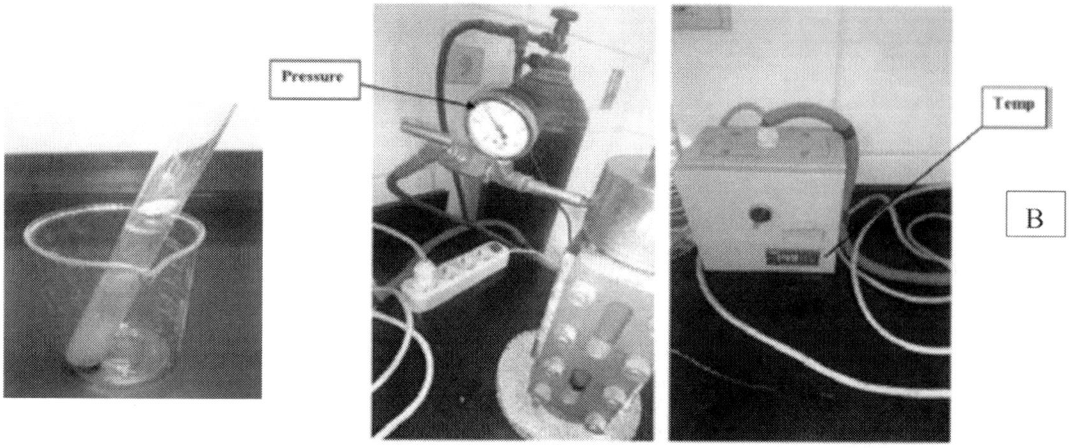

Fig. 2- Solubility test equipment and procedures, a) at room conditions (77°F, and 14.7 psi), b) at (300°F and 1000 psi).

Corrosion tests were performed to examine the effect of the new chemical on the tubing at the downhole conditions. The same corrosion inhibitor was used as that was decided to be used during the scale removal job operations. The tests have been carried out on a received piece from carbon steel pipe (L-80) used in this job. After surface cleaning of the pieces (15 cm^2), it was immersed in Teflon tubes containing 100 cm^3 solvent. The tubes were put in autoclaves and subjected to the test temperature (300°F) for 16hr. At the end of the test period, the steel samples were cleaned, dried and weighed. Table 4 shows the corrosion tests results in corrosion inhibitor concentrations of 1 and 1.5 vol%.

Corrosion Inhibitor Concentration, vol%	Weight Loss, lb/ft^2	Test Period, hr	Temperature, ^{0}F
1	0.0395	16	300
1.5	0.0385	16	300

Table 4- Corrosion inhibitor measurements at different corrosion inhibitor concentrations.

Well Treatment

This treatment aims to remove the scale deposits accumulation from downhole and clean the perforations to allow matrix accessibility, enhance productivity, restore production, and avoid by-products. The descaling operation has been performed on two stages; wellbore cleaning stage and well stimulation stage. Table 5 shows the cleaning chemicals used to clean the wellbore to allow matrix treatment. Wellhead pressure was maintained at 300 psi

Solution	Chemical	Volume, gal
Slickwater	Freshwater	12,600
	Ammonium Chloride	4,200
	Foaming agent	76
	Water Gelling agent	600
	Surfactant	26

Table 5- Wellbore cleaning chemicals.

Matrix stimulation stage aimed to enhance formation productivity prior to restoring production. Four different stages were injected; preflush (Diesel, MS, and Xylene), main treatment (scale removing agent soaking for 16 hrs.), Over-Flush (NH_4Cl, mutual solvent), and Flow back. Table 5 shows the chemicals used in the main treatment.

Solution	Chemical	Volume, gal
CT Pickling Solution	HCl	106
	Surfactant	1
	Corrosion Inhibitor	2
	Freshwater	100
Preflush fluid	Freshwater	8,090
	NH_4Cl	3,500 lb
	Surfactant	17
Pot Solution	Scale removing agent	950
Main Treatment	Scale removing agent	5,300
Post Flush	Fresh Water	5,590
	NH_4Cl	2,660 lb
	Mutual Solvent	636
	Clay Stabilizer	13

Table 5- Main treatment chemicals.

Results

The job has been completed safely and efficiently, and the well started flowing naturally and with 10 times its production before treatment. **Fig. 3** shows the production enhancement after the good treatment to 1390 bbl/day comparing to 133 bbl/day before the treatment and the productivity index increased to 0.5-0.6 bbl/day/psi.

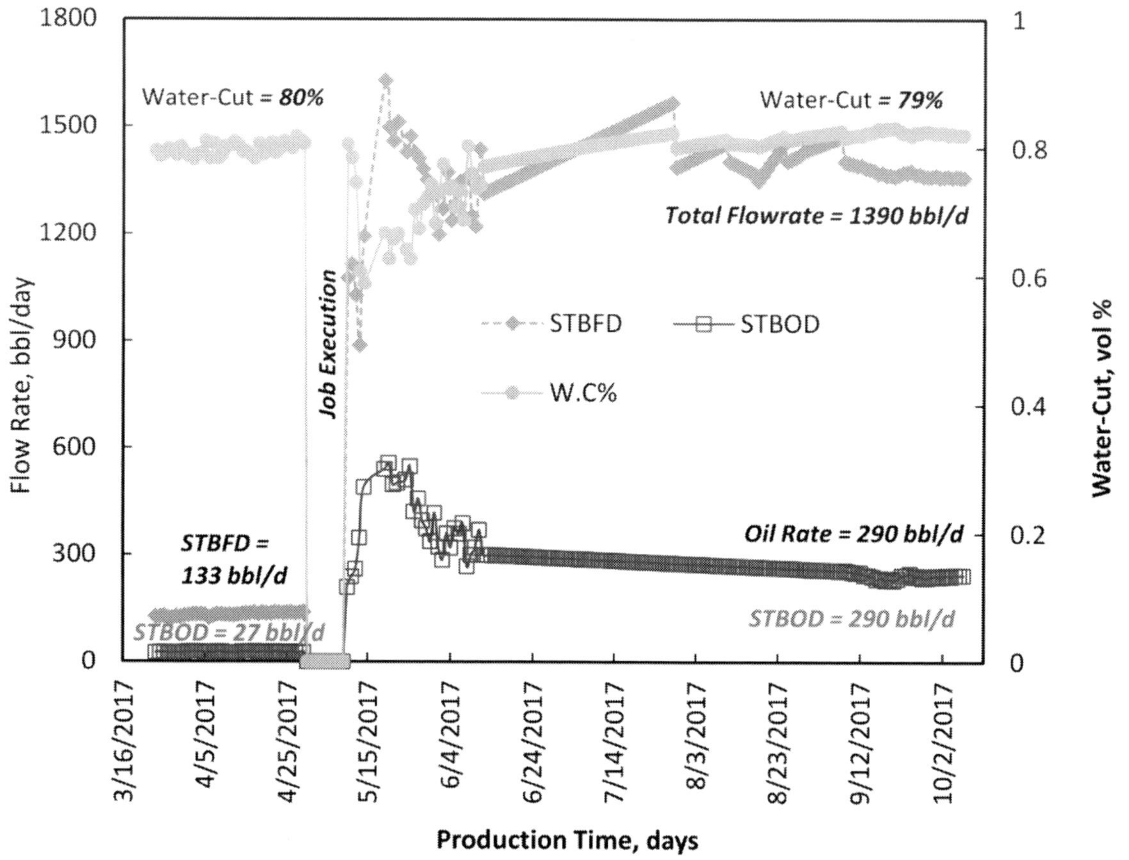

Fig. 3 –Total production and water-cut before and after the treatment.

Conclusions

This paper presents the using of new iron sulfide scale removing agent. The experimental study was conducted to evaluate this chemical then it was applied on actual field case. The main conclusions are summarized as follows;

1. The new chemical was able to dissolve up to 77% of the iron sulfide scale sample at room temperature and downhole conditions of 300°F, and 1000 psi.
2. The chemical is non- corrosive to the tubing material (L-80) at downhole conditions.
3. The field case shows increasing on the productivity index 10 times its initial value by dissolving the iron sulfide scale.

The new chemical is non- corrosive dissolver for iron sulfide scale deposits.

Acknowledge

The authors thank SAPE SAHARA Chemicals Solutions (SCS) SCO and East Zeit Petroleum companies for their permeation to publish this paper.

References

Bajammal, F.A., Biyanni, H.M., Riksa, A.P. et al. 2013. Scale Management in Mature Gas Field: Case Study of Peciko. Presented at the International Petroleum Technology Conference, Beijing, China, 26-28 March. IPTC-16675-MS. https://doi.org/10.2523/IPTC-16675-MS.

Farrokhrouz, M. and Asef, M.R. 2010. Production Enhancement Via Scale Removal in Nar Formation. Presented at the SPE Production and Operations Conference and Exhibition, Tunis, Tunisia, 8-10 June. SPE-135854-MS. https://doi.org/10.2118/135854-MS.

Ford, W.G.F., Walker, M.L., Halterman, M.P. et al. 1992. Removing a Typical Iron Sulfide Scale: The Scientific Approach. Presented at the SPE Rocky Mountain Regional Meeting, Casper, Wyoming, 18-21 May. SPE-24327-MS. https://doi.org/10.2118/24327-MS.

Gamal, M., Raafat, M., Thabt, M. et al. 2013. Integrated Reservoir Study of Mature Oil Field Resulted in Rejuvenation of an Abandoned Oil Reservoir to Add New Reserves and Production, Gas - Egypt. Presented at the North Africa Technical Conference and Exhibition, Cairo, Egypt, 15-17 April. SPE-164760-MS. https://doi.org/10.2118/164760-MS.

Gomes, R., Mackay, E.J., Deucher, R.H. et al. 2012. Impact of Reservoir Reactions on Thermodynamic Scale Predictions. Presented at the SPE International Conference on Oilfield Scale, Aberdeen, UK, 30-31 May. SPE-155255-MS. https://doi.org/10.2118/155255-MS.

Leal Jauregui, J.A., Solares, J.R., Nasr-El-Din, H.A. et al. 2007. A Systematic Approach to Remove Iron Sulphide Scale: A Case History. Presented at the SPE Middle East Oil and Gas Show and Conference, Manama, Bahrain, 11-14 March. SPE-105607-MS. https://doi.org/10.2118/105607-MS.

Mackay, E.J. 2003. Modeling in-Situ Scale Deposition: The Impact of Reservoir and Well Geometries and Kinetic Reaction Rates. *SPE Prod & Fac* **18** (01): 45-56. DOI: https://doi.org/10.2118/81830-PA

Nasr-El-Din, H.A. and Al-Humaidan, A.Y. 2001. Iron Sulfide Scale: Formation, Removal and Prevention. Presented at the International Symposium on Oilfield Scale, Aberdeen, United Kingdom, 30-31 January. SPE-68315-MS. https://doi.org/10.2118/68315-MS.

Nasr-El-Din, H.A., Fadhel, B.A., Al-Humaidan, A.Y. et al. 2000. An Experimental Study of Removing Iron Sulfide Scale from Well Tubulars. Presented at the International Symposium on Oilfield Scale, Aberdeen, United Kingdom, 26-27 January. SPE-60205-MS. https://doi.org/10.2118/60205-MS.

Quintero, L., Felipe, M.J., Miller, K. et al. 2018. Microemulsions Increase Well Productivity by Removing Organic Deposits and Inorganic Scale in One Step. Presented at the SPE International Conference and Exhibition on Formation Damage Control, Lafayette, Louisiana, USA, 7-9 February. SPE-189514-MS. https://doi.org/10.2118/189514-MS.

Yap, J., Fuller, M.J., Schafer, L. et al. 2010. Removing Iron Sulfide Scale: A Novel Approach. Presented at the Abu Dhabi International Petroleum Exhibition and Conference, Abu Dhabi, UAE, 1-4 November. SPE-138520-MS. https://doi.org/10.2118/138520-MS.

522817 A Deep Neural Net Model for Oil/Water Separation in Oil Production Pipelines

Abstract:

High fidelity 3-D engineering simulations are valuable tools providing recommendations and solutions difficult to obtain otherwise. However, those simulations can be costly due to expensive computational requirements. In this study, a deep neural network model trained by computational fluid dynamics simulations capable of predicting oi/water separation in horizontal oil pipelines was developed to accelerate and replace the CFD simulations. The goal is to provide useful information for oil production operators to effectively determine the use of corrosion inhibitors, which can add significant cost to the operation, and incorrect decisions can have serious impact on our environment.

The CFD model was developed in the past 4 years for oil/water separation in horizontal pipelines to predict the water wetting probability on carbon steel pipeline surfaces [1]. The model has been shown to correctly reproduce the flow regime map widely used in industry across ranges of water cut, mixture velocity, pipe diameter, oil density, viscosity and surface tension. However, the model requires expert set up and can take days to complete. Recently deep learning neural network algorithms has gaining momentum in the way that it automates the correlation construction between input and output data and allows complex hyper-dimensional correlations to be built with high accuracy. The efficiency comes from the optimization procedure unique to convolutional neural network (CNN) models. By parametrizing the CFD simulations it is possible to pre-calculate the results in all the necessary parametric spaces and use them to train the CNN model.

The preliminary results in two parametric spaces, water cut and mixture velocity, are presented. The trained CNN model showed very high accuracy with the mean square errors lower than 0.01 in all cases. A total of ~4,000 data points were used to train the CNN model. The training time was around 4 minutes, and the inferencing time was less than 5 milliseconds, a performance gain of 106 in comparison to the CFD simulations.
The model is being extended to include the other parameters aforementioned. It can also be extended to model other types of predictions as well, such as erosion and corrosion, and with more complex geometries other than straight pipes. The possibilities are endless.

[1] K. Tsai, "Modeling the separation of oil and water in pipelines," Paper 452a, 2017 AIChE Annual Meeting, Minneapolis, USA.

References

Hu, H. and Y.F. Cheng, "Modeling by computational fluid dynamics simulation of pipeline corrosion in CO_2-containing oil-water two phase flow," J. Petroleum Sci and Eng., **146**, pp. 134-141, 2016.

Kanwar, S., *Study and modeling of sweet corrosion of multiphase mixtures in horizontal pipelines*, Master of Science Thesis, Ohio University, USA, 1994.

Tsai, K., "Corrosion modeling using electrochemistry and computational fluid dynamics," Paper No. 355b, 2017 AIChE Annual Meeting, Minneapolis, MN, USA., Oct., 2017a

Tsai, K,, "Modeling the separation of oil and water in pipelines," Paper No. 452a, 2017 AIChE Annual Meeting, Minneapolis, MN, USA, Oct. 2017b

Zhang, G.A. and Y. F. Cheng, "Electrochemical characterization and computational fluid dynamics simulation of flow-accelerated corrosion of X65 steel in a CO_2-saturated oilfield formation water," Corros. Sci. 52, pp. 2716-2724, 2010.

530558 Hydrate Risk Management during Cold Restart Operation Using MEG and Khi

Abstract:

Hydrate blockage formation in offshore flowlines may induce production stoppage and operational hazards. Although there have been numerous application of KHIs and AAs in offshore oil and gas fields, MEG is still a reliable solution to avoid hydrate blockage formation in offshore flowlines. The required concentration of MEG in aqueous phase to avoid hydrate blockage may be more than 50 wt% in deep water or arctic oil and gas fields, therefore a reduction in the MEG concentration can result in significant investment saving. Our previous works suggested that the growth rate of hydrates can be controlled in the MEG concentration lower that the value required to fully avoid the hydrate formation, so called under-inhibition with MEG.

This work investigates the effect of adding the hydrate inhibitors on hydrate blockage formation during cold restart operation while mixing the fluids at constant stirring rates of 200, 400, and 600 rpm. Depending on the mixing rate, liquid phase became stratified (200 rpm), partial dispersing (400 rpm), and full dispersing (600 rpm). A high pressure autoclave was used in this study, which is equipped with a magnetic stirrer coupling and a two-blade impeller. A total liquid volume of 200 mL was loaded into the autoclave cell which had an internal volume of 510 mL. The cell was immersed in a temperature-controlled liquid bath connected to an external refrigerated heater. A platinum resistance thermometer monitored the temperature of the liquid phase inside of the autoclave with an uncertainty of 0.15 °C. The pressure was measured by a pressure transducer with an uncertainty of 0.1 bar in a range of 0 - 200 bar. To provide vigorous mixing of the liquid phase, a two-blade impeller on a solid shaft coupled with the motor was used. The impeller was located in the bottom of the shaft. A torque sensor with platinum coated connector measured the torque of continuously rotating shaft with an uncertainty of 0.3%. It used a strain gauge applied to a rotating shaft and a slip ring that provides the power to excite the strain gauge bridge and transfer the torque signal. To safely operate the motor and torque sensor, safety-lock was implemented in monitoring/control system to stop the motor once the torque value became higher than 50 N cm during the hydrate formation process.

For water and decane mixture without hydrate inhibitors, hydrates formed instantly upon mixing and leads to the hydrate blockage within 13.9, 18.8, and 42.2 min for stratified, partial dispersing, and full dispersing liquid phase, respectively. The resistance-to-flow was estimated from the measurement of torque changes during the hydrate formation. Sever torque spikes were observed for the water and decane mixture without hydrate inhibitors. The hydrate growth rate was highest initially, but decreases with increasing hydrate fraction in liquid phase. When it reached the steady value, torque start to spike possibly due to agglomeration and bedding of hydrate particles. Adding 20 wt% mono-ethylene glycol (MEG) to the water phase could suppress the torque increase while hydrate formation proceeds to the final fraction, suggesting MEG may prevent the agglomeration and bedding of hydrate particles for all flow regimes. However its performance decreased at 10 wt% MEG concentration, where surge of relative torque was observed although the resistance-to-flow was not enough to stop the fluids flow. The addition of 1.0 wt% Luvicap suppressed the hydrate onset for 155.0 min at mixing rate of 200 rpm, however soon lost its efficacy with increasing mixing rate to 400 and 600 rpm. The growth rate was also maintained low until hydrate fraction reached about 0.07, then catastrophic hydrate growth was occurred followed by increasing torque increase. These results suggested that the hydrate formation mechanism during cold restart would be highly dictated by the mixing rate and corresponding flow regime, thus appropriate hydrate inhibition strategy must be developed to manage its risk. In this work, under-inhibition with MEG, adding Luvicap, and hybrid inhibition with MEG and Luvicap were tested. Among them, addition of 20 wt% MEG to the water phase was the best to manage the hydrate blockage risk. Luvicap was effective at stratified condition, however may induce blockage at high degree of turbulence. For MEG and Luvicap solution, hydrate formation initiated in interface between water and decane, then proceeded to gas phase without affecting the resistance-to-flow. The performance of hydrate inhibitors must be evaluated based on relevant data measurement and visual observation to better describe the hydrate formation mechanism.

536687 A Microfluidics Based Study on the Effect of Immiscible Huff-n-Puff Process on Residual Oil Saturation in Hydrophilic and Hydrophobic Porous Media

Abstract:

Cyclic gas injection or "Huff-n-Puff" method is a widely applied enhanced oil recovery (EOR) method in depleted reservoirs. It extracts additional oil by reducing oil viscosity from gas diffusions. In this work, microfluidic experiments were conducted to investigate the performance of immiscible "Huff-n-Puff" process using a novel laboratory setup. Residual oil saturation (S_{or}) in each Huff-n-Puff cycle was studied after water and gas flooding, together with the effects of aspect ratio and wettability.

Uniform pore network chips with hydrophilic and hydrophobic wetting surfaces and aspect ratios ranging from 1.97 to 23.94 were used. In this study, the aspect ratio was defined as the ratio of the maximum cross-sectional area of the pore and that of the throat, both measured perpendicular to the flow direction. Water, n-decane, and CH_4 were used as aqueous, oil, and gas phases, respectively. The aqueous and oil phases were dyed with methyl blue and Sudan red, separately to increase visual contrast for enhanced fluid saturation measurements.

In each experiment, a microfluidics chip was first saturated with water then oil to achieve the initial saturations. It was then flooded with water and gas to achieve residual fluid saturations before pressure reduction steps in "Huff-n-Puff". Water and gas injection rates were adjusted to maintain capillary numbers in the range of 10^{-7} to 10^{-5}, to match values of reservoirs. A constant back pressure of 100 psig was maintained before "Huff-n-Puff" started. Four pressure reduction steps were performed by reducing the back pressure by 25 psig in each step. Chip and pore scale images were taken, and analyzed using ImageJ software to measure the residual fluid saturations.

For the continuous gas flooding prior to "Huff-n-Puff", the results show that the oil saturation increases semi-logarithmically with aspect ratio (log AR vs. S_{org}) for both hydrophilic and hydrophobic chips. During the four "Huff-n-Puff" steps, up to 20 %PV oil was produced in hydrophilic chips, whereas less than 5 %PV of oil was produced in hydrophobic chips.

537532 Estimating the Drainage Area of Frac-HIT or RE-Fractured Horizontal Well

Abstract:

Production from unconventional (low-permeability) plays has emerged as a major source of oil and gas production in North America in the past decade. Use of horizontal wells stimulated by multiple transverse hydraulic fractures has enabled this development. To extract the most amount of hydrocarbons from a given acreage, it has become common practice to space multiple horizontal wells very close to each other. This often results in so-called "frac-hits", i.e. a phenomenon where production from an existing well is considerably decreased due to interaction with a new adjacent well being fractured, as fractures from the new well may reach the original well's drainage area. Production from an existing well may also not be optimal due to the original fracturing resulting in limited drainage area or due to reduction of the original fractures' conductivity over time. This makes it appealing to re-stimulate the well via "re-fracturing" operations, to increase well productivity. For both cases – frac-hits or re-fracturing - it is important to quantify the evolution of the drainage area of such existing wells. Developing a model that accomplishes this calculation is the focus of this presentation. The objective is to develop a relatively simple methodology that can easily be used in the field. The approach taken is as follows:

- Develop computer simulations of scenarios that model drainage area variations due to frac-hits or re-fractures,
- Conduct rate-normalized pressure data analysis to look for factors that control the alteration of production,
- Predict simple algebraic relationships/equations with the help of identified factors to quantitatively predict altered performance of a fractured horizontal well,
- Verify the predicted equations by application to data from real wells.

The approach presented above attempts to combine the advantages of sophisticated simulation employing fundamentals and of simplified algebraic equations collectively described as decline curves.

515235 Multiphase Flowloop Investigation of Transportability and Flow Properties of Highly Concentrated Hydrate Slurries

Abstract:

Hydrate formation, if not properly monitored and controlled, may lead to pipeline blockage. In order to avoid pipeline blockage and other hydrate formation risks, chemical additives are added to the system. Additives such as anti-agglomerants (AA) help improve hydrate transportability by dispersing the formed hydrates into slurries and preventing them from sticking to the pipe wall. This enables transportation of highly concentrated slurries. However, the high hydrate volume fractions (HVF) slurries may exhibit complex rheology. There is therefore a great need to correlate flow properties such as friction factor and viscosity to HVF. Hydrate slurry transport is important whether hydrates are deliberately generated for energy storage purposes or hydrates formed because of the prevailing flow conditions.

In this work, a multiphase flowloop was use to generate gas hydrate slurries with models oils of different viscosities. Tests were performed at high liquid loading (+75%) and at intermediate watercuts (30 to 70%). Pressure drop was monitored with increasing HVF. AA dosage was maintained at the recommended dosage, i.e., about 2 vol% with respect to aqueous phase for the AA used.

The first part of the manuscript presents viscosity predictions using the non-Newtonian multiphase friction factor model developed by Garcia et al. 2003. Results indicate that viscosity increases almost linearly with increasing HVF up to 18% HVF but exponentially above 25%HVF. This behavior may suggest onset of heterogeneous flow, i.e., AA fails at high HVF.

The second part of the paper presents a comparative study of viscosity models including those developed on benchtop scale. It was observed that models that account for carrier fluid viscosity in predicting slurry viscosity agreed better with experimental data. Presence of free gas phase presented the greatest challenge to viscosity and friction factor models.

Lastly, transportability assessment showed that low viscosity systems were greatly affected by increase in HVF whereas high viscosity systems in addition to HVF, they were significantly affected by temperature changes. Low viscosity flow systems exhibited heterogeneous flow behavior and even plugging at higher HVF. On the other hand, high viscosity oils formed emulsion and were characterized by high shear heat buildup.

530166 Mitigation of Severe Slugging with Internal Model Control

Abstract:

In the offshore oil production field, unstable fluid flow may occur when two-phase fluid passes through a vertical riser connected to a downwardly inclined subsea line, or when a pipe is laid in a hilly terrain. This is called severe slugging or terrain induced slugging. As it passes through this section, liquid accumulates at the bottom of the pipeline and begins to block the passage through which the gas phase can pass, compressing the gas until it is pressurized to move the liquid. When the gas is able to overcome the gravitational head of the liquid, the accumulated liquid will come out at once. Mitigation of severe slugging is essential when designing an offshore plant because the liquid level of the separator may suddenly increase or vortex induced vibration (VIV) may appear in the riser due to the severe slugging.

The most common way to eliminate severe slugging is choking the topside valve. If the valve is choked under certain valve opening at the topside separator, back pressure and the velocity at the choke increases. This makes the liquid holdup to increase in the riser because more gas phase stream passes through valve than liquid phase stream. Positive perturbation of liquid holdup increases weight in riser and causes the liquid to fall down so that the pressure drop over the riser grows. Increased pressure drop increases the gas flow, allowing the gas to flush the liquid well and eliminating severe slugging.

Therefore, in this study, a multiphase flow loop which is consisted of downwardly inclined pipe and vertical riser in lab scale was designed and installed in order to generate and mitigate severe slugging. This flow loop consists of a vertical line of about 5.7m, a horizontal line of 12m, and a choke valve is installed on the upper part of the vertical line to mitigate the severe slugging. To compare the experimental results, model was made by inputting the dimensions of the flow loop using OLGA, a multiphase flow simulation program.

First, the opening of the topside valve was manually controlled to confirm that severe slugging was mitigated. A bifurcation diagram was made to more clearly identify which flow this system has at each valve opening. Bifurcation diagrams have been used to plot the values of pressure versus the values of valve opening for the slugging system. In the stable regions the plot consists of a unit curve showing the exact value of the pressure in simulations or the average of very small pressure oscillations in experiments. In unstable regions the plot consists of three curves, one for steady state conditions and the two others showing the maximum and minimum of oscillations at each value of valve opening over the work range of choke valve. Valve opening at the point where the severe slugging flow changes to stable flow is expressed as critical valve opening. Experimental results show that severe slugging disappears at the opening below the critical valve opening, and that various critical valve opening values ranging from 0.3 to 0.7 are obtained depending on the liquid and gas flow rates.

The OLGA model was tuned by adjusting the valve coefficient or the air tank diameter so that a bifurcation diagram similar to that of the experiment was obtained. Then, the model-based control logic was constructed by estimating the transfer function using the OLGA model. When estimating the transfer function, the inlet gas buffer pressure data near the critical valve opening were used and the System Identification Application provided by MATLAB was used. This application estimates the transfer function by numerical method based on the data calculated using OLGA. The transfer function is estimated to have two poles and one zero, and data from the OLGA data and the estimated transfer function show a high agreement rate of 80-90%.

The estimated transfer function is used to estimate the gain of the PI controller, using the method called the Internal Model Control (IMC). The IMC method is one of the model predictive control methods, which minimizes the difference between the transfer function of the actual system and the transfer function of the predicted model. The control gain is calculated by using this method and the calculated gain is input to the PI controller of the experimental equipment.

As a result, it was observed that stable flow can be obtained even if the valve opening is increased up to 0.45 from 0.33, which is the critical valve opening of Usg=0.7m/s, Usl=0.1 m/s condition when manual choking is performed. Likewise, it was also observed that stable flow can be obtained even if the valve opening is increased up to 0.9 from 0.65, which is the critical valve opening of Usg = 0.8m/s, Usl=0.2 m/s condition when manual choking is performed. When the PI controller gain is obtained through the transfer function of the flow loop system as described above, mitigation is possible even at the valve opening which is larger than manual choking. Therefore, it can be said that it is a very efficient control method because it can provide more production to topside while controlling severe slugging than manual choking.

525111 Investigation of the Interaction between Wax Precipitation and Hydrate Formation in Water-in-Oil (W/O) Emulsions

Abstract:

The hydrate blockage and wax deposition may exist simultaneously in the transport system, which will pose a great threat to the safety of pipelines and enhance the safety hazards of subsea production system. Therefore, the study of interaction between wax precipitation and hydrate formation in water-in-oil (W/O) emulsions, is of great significance for the security of development in deep-water waxy oil and gas fields. Experiments of natural gas hydrate formation in W/O emulsions containing wax crystals were carried out in the high-pressure autoclave. The macro-parametric data such as pressure, temperature, hydrate induction time, hydrate formation amount and rate was compared and analyzed. Besides, the effect of hydrate formation and dissociation on wax precipitation was studied combined with the microscopic data analysis obtained from the polarized light microscopic observation. The results indicated that wax crystals had no influence on the stage behavior of natural gas hydrate formation process. Wax crystals were absorbed on the surface of water droplets under the synergistic effect of surfactant, increasing the mass transfer resistance of hydrate nucleation, hence prolonging the hydrate nucleation induction time. An empirical model for predicting hydrate nucleation induction time in W/O emulsion systems containing wax crystals was developed based on the Freundlich adsorption isotherm. The precipitated wax crystals in W/O emulsions would affect the structure of the hydrate shell, causing gas hydrate to grow at a lower rate, but the total hydrate formation amount was increased compared to the emulsified systems without wax crystals. More wax crystals precipitated in the systems after the formation and dissociation of natural gas hydrate compared to the systems without hydrate formation. The box dimension of wax crystals was relatively larger due to hydrate formation and dissociation, implying that the structure of precipitated wax crystals was more intricate. The findings of this work provide a preliminary insight into the interaction behavior of hydrate and wax, which is an on-going research topic of flow assurance in deep-water offshore fields.

527622 Electrical Treatment of Waxy Crude Oil to Address Wax-Related Flow Assurance Issues

Abstract:

Waxy crude oils account for a significant portion of petroleum reserves. The wax precipitation from waxy crude oils caused by temperature reduction may manifest into several wax-related flow assurance issues including enhanced viscosity, wax deposition, or gelled pipeline. In the present work, we demonstrate a novel method of electrical treatment that shows great promises in addressing these issues. First, the low-temperature viscosity of waxy crude oil could be significantly reduced by treating it with high-voltage direct electric field. A highest viscosity reduction of 70% was obtained after the crude oil was electrically treated at 0.8 kV/mm for 90 s. The energy consumption of the electrical treatment was estimated to be less than 1% of that of the conventional heating method to achieve the same viscosity reduction performance. Interestingly, the heavy and polar components of crude oil such as asphaltenes and resins are found to undermine the performance of electrical treatment. Moreover, electrically treating waxy crude oil below the wax appearance temperature can also significantly reduce the overall strength of the formed waxy gel, which might facilitate the restart of a gelled pipeline. The results from microscopic observations indicated that the polarization and aggregation of wax crystals in the electric field might be responsible for the observed effects of electrical treatment. These findings suggest that electrical treatment may serve as a promising and energy-efficient technique to address wax-related flow assurance issues.

543226 CO_2Â Foam Stabilization using Zwitterionic and Nonionic Surfactants

Abstract:

Selection of surfactant for enhanced oil recovery and other upstream applications is a challenging task. For upstream foaming applications, it should be thermally stable, compatible with reservoir brine and other chemicals, have lower adsorption on reservoir rock, has high foamability and foam stability, and should be economically viable. In this work, Foamability and foam stability of three different surfactants were evaluated using dynamic foam analyzer. These surfactants were fluorinated nonionic, fluorinated zwitterionic, and hydrocarbon zwitterionic surfactants. Effect of surfactant type and structure, concentration, temperature, polymer additions, salt addition, and type of injected gas was investigated on foamability and foam stability. The foamability was assessed using the volume of foam produced by injecting a constant volume of gas and foam stability was determined by half-life time. The maximum foam generation was obtained using hydrocarbon zwitterionic surfactant. However, the foam generated using fluorinated zwitterionic surfactant was more stable. A mixture of solution containing zwitterionic fluorinated and hydrocarbon fluorinated surfactant showed better foam generation and foam stability. The foam generated using CO_2 has less stability compared to the foam generated using air injection. Addition of polymer significantly improved the foam stability and have no effect on foam generation. Presence of salts increases the foam stability and foam generation. This study helps in optimizing the conditions of foam injections for enhanced oil recovery.

541865 Evaluation of Wax Precipitation Behavior of Wax Deposit: The Effect of Oil Flow Condition

Abstract:

Research Interests: Petroleum Storage and Transportation Engineering

Teaching Interests: Petroleum Storage and Transportation Engineering

EVALUATION OF WAX PRECIPITATION BEHAVIOR OF WAX DEPOSIT: THE EFFECT OF OIL FLOW CONDITION

Xuedong Gao[1], Qiyu Huang[1,*], Yijie Ren[2], Weidong Li[1], Xue Dong[1]

[1] Beijing Key Laboratory of Urban Oil and Gas Distribution Technology, China University of Petroleum-Beijing, Beijing, China

[2] Shanghai Jiao Tong University, Shanghai, China

ABSTRACT

Wax deposition is a major flow assurance challenge in waxy crude oil production and transportation. Wax precipitation behavior, playing some important role in wax strength, is considered to be a decisive factor that affects wax deposit property and further determines pipeline pigging. It is intrinsically correlated to the oil flow condition. This paper aims to provide some qualitative and quantitative insights into the relationship between oil flow condition and wax precipitation behavior of wax deposit. A laboratory-scale flow loop apparatus was employed. Four sets of crude oils collected from field were used. Variations of wax content and wax appearance temperature (WAT) of wax deposit against shear stress, radial temperature gradient and radial wax molecule concentration gradient near the pipe wall were well studied.

Differential scanning calorimetry (DSC) tests on wax deposits indicate that higher shear stress and radial temperature gradient ratchet up the wax content and WAT and down the wax molecule concentration gradient. Based on this, empirical formulas for predicting the wax content and WAT were developed. Verification experiments show that for wax content, the maximum and average relative deviations are 21.5% and 13.2%, respectively. As for WAT, the maximum and average absolute deviations are 5.4°C and 3.6 °C, respectively.

By investigating the variations of wax content and WAT of wax deposit against shear stress, radial temperature gradient and radial wax molecule concentration gradient, this paper promotes the understanding of how oil flow condition affects the wax precipitation behavior of wax deposit. Pipeline pigging can also enjoy some benefits from this work.

Keywords: wax deposition; wax precipitation behavior; oil flow condition

(A) (B)

Figure 1. Effect of wall shear stress on wax content and WAT of wax deposits: (A) wax content and (B) WAT.

(A) (B)

Figure 2. Effect of radial temperature gradient on wax content and WAT of wax deposits: (A) wax content and (B) WAT.

(A) (B)

Figure 3. Effect of radial wax molecule concentration gradient on wax content and WAT of wax deposits: (A) wax content and (B) WAT.

Table 1. Verification results for the empirical formulas

Number	Wax content (%)			WAT (°C)		
	Measured value	Calculated value	Relative deviation	Measured value	Calculated value	Absolute deviation
1	39.6	34.7	-12.4	55.2	51.8	-3.4
2	41.9	46.3	10.5	56.0	58.8	2.8
3	20.9	25.4	21.5	48.4	53.8	5.4
4	21.6	20.9	-3.2	45.7	49.5	3.8
5	38.9	42.5	9.3	53.3	58.6	5.3
6	66.2	58.0	-12.4	63.1	66.7	3.6
7	26.6	30.1	13.2	53.4	55.5	2.1
8	41.2	35.5	-13.8	65.	61.3	-3.7
9	40.9	48.4	18.3	64.9	68.3	3.4
10	46.8	38.6	-17.5	58.7	56.4	-2.3

542620 The Impact of Shale Gas and Oil on the Chemical Industry

Abstract:

Lower alkanes and aromatics are very useful as both fuels and chemical raw materials. Occasionally supply disruptions constrained availability but also encouraged investigation of alternative feedstocks and chemistries. Recent new technologies for the production of unconventional gas and oil sources from tight shale formations significantly increased availability of both methane and condensate, radically altered process economics, resulted in the construction of much new intermediates capacity, and multiplied the ways traditional and alternative chemistries might be exploited. This presentation will outline historically how natural gas components became useful as chemical feedstocks, how government regulations occasionally led to perturbations in availability and price, how supply crises led to the development of alternative chemistries, catalysts, and ultimately the exploitation of shale resources. More specifically, will be discussed how excess coproducts, especially condensate, from increasing shale gas and oil production for energy are having significant unexpected impacts on the chemical process industries and opportunities for new process chemistry research.

541186 Sustainable Energy and Chemicals: Past, Present, and Future

Abstract:

The last 40 years has seen the U.S. energy situation vacillate from gasoline rationing and concerns over Peak Oil, to projected energy and feedstock prices that were noncompetitive in global markets for chemicals production, to the economic revolution of frack gas and light tight oil. The importance of low cost energy and feedstocks in driving economic opportunity will be discussed relative to this history, along with the importance of sustainability in decision making to gain stakeholder buy-in and acceptance. Strategies for managing career paths and competencies despite changing targets and high uncertainty will be addressed.

542618 Disruptions: What the Future May Hold

Abstract:

Disruptive forces are all around us. They change the way we communicate and work. They change the way we develop goods and services. They also change the types of products that are manufactured. Examples such as the way in which we produce and consume energy or the feedstocks we use to produce product have seen many changes over the course of the last few decades, and these disruptions are expected to continue. The chemical engineering discipline and the expectations for a chemical engineerâ€™s skills has also changed and adapted as a result of these disruptive forces. A discussion of the broad themes associated with these forces and the implications that they may have to chemical engineers in the future will be discussed.

542621 Geopolitical Factors Influencing the Evolution of the Chemical Industry

Abstract:

The basic chemicals and plastics industry is built on a simple principle: find a cheap feedstock and convert it into something more valuable. However, the right feedstock depends on place and time. Certain policies and actions of various governments have influenced the evolution and profitability of the chemical industry during the past 50 years, sometimes in surprising ways. For example, high oil prices and US government funding encouraged the development and commercialization of coal gasification technology during the late 1970s and early 1980s. This enabled China to develop a competitive basic chemical industry using coal as a feedstock during the last 25 years. Similarly, an abrupt increase in oil prices during the first decade of the 21st century encouraged the present shale gas revolution, causing a major shift in the economy of producing basic chemicals and polymers in the US. If history is our guide, then what governments choose to fund and regulate may be as important as market forces in determining the shape of our industry and profession in the future.

542622 Agility & Resilience: How to Maintain Career Competitiveness in the Changing Chemical Industry

Abstract:

The world around is changing. The shift to protectionism, global instability and price volatility of raw materials and feedstocks has the potential to wreak havoc on the chemical industry. With the revolution of shale gas in North America and the sudden drop in feedstock costs, the chemical sector in the US has reasserted its historic dominance. Learn how the role of a chemical engineer can contribute to industry growth and resilience, while adapting to allow for personal agility throughout the career life cycle.

541565 25 by 25: Chemical Engineering in the Next 25 Years

Abstract:

In 2008, the centennial year for AIChE, we asked 25 chemical engineers for their visions of the profession from around the US and the world: Nine faculty, eight from industry, and eight post-docs and graduate students. We posed four questions related to each individual's industrial sector or area of research, asking for extrapolation, new impacts on existing sectors, new sectors, and an optional, more open-ended comment on the future of the profession as a whole.

In this 110th year of AIChE, we have revisited these questions with many of the same people. Ten years has seen many developments in our profession, in society, and in our thought-leaders' experiences. Here we report on their perspectives, comparing them to the 2008 visions.

541542 The Future of Chemical Engineering Itself

Abstract:

Chemical engineeringâ€™s future is bright because it is vital to so much of society and industry -- and its future is complicated by its rapidly expanding universe of activities. It must represent depth in core skills centered on the chemical sciences including biology and materials. At the same time, it must be transdisciplinary; it must exemplify the need and value of reaching across disciplines in borderless ways.

One way is to create and disseminate a compelling vision and strategy for chemical engineering, research, and education for the future. AIChE and the National Academies are working together to that end. In 1988, the National Academies published "Frontiers in Chemical Engineering: Research Needs and Opportunities.â€ Its committee of authors outlined a wide-ranging roadmap to turn promising research opportunities into reality, while calling for researchers to embrace new frontiers.

Since 1988, that has happened -- along with many other transformative changes like the Web, fracking, personalized medicine, and social media. A new nationwide effort, aided by international contributions, will address both the identity of chemical engineering and its future directions. Its goal is to ensure that the people in chemical engineering are primed to meet society's needs and rapid changes.

544708 Accelerating Innovation through Academic-Industrial Partnerships

Abstract:

In 2011, The Dow Chemical Company launched the University Partnership Initiative (UPI) to sponsor collaborative research at leading U.S. Universities. This 10-year commitment aims to address fundamental questions in industrial chemistry and chemical engineering with alignment to Dowâ€™s next-generation process technology and product development efforts. To date, this portfolio of projects has generated an impressive array of intellectual property, publications, and numerous examples of the collaborative developments implemented at Dow. This presentation will introduce UPI program and highlight several contributions from our collaborators and leaders in the chemical engineering profession.

540559 Maximizing Uptime, Efficiency, and Safety of Industrial Operations through Early Risk Detection

Abstract:

Achieving zero incidents while maximizing process uptime and efficiency is a key vision for operating companies. An important building block towards this vision is establishment of a proactive risk mitigation culture, supported by effective systems and a strong management. Studies have shown that most of the incidents and unexpected process failures can be avoided if the operating teams get timely information about developing risks and take preventive actions early on. While advances have been made in process monitoring and automation in the past decades, there still remains significant technological and behavioral gaps that prevent plant operations from proactively mitigating process risks.

A modern industrial plant monitors thousands of parameters, generating upwards of 50-100 million data points every day. Thanks to recent breakthroughs in machine learning and artificial intelligence approaches, today there are autonomous systems that can sift through this data and point out meaningful and timely insight. This can help operating teams ascertain process issues that are hidden in the data, long before process variables reach critical levels.

In this presentation, a new autonomous system, Dynamic Risk Analyzer™ (DRA), will be introduced that points out early indicators of process issues using proprietary machine learning. The risk indicators enable operations team to respond and make required process changes, days and sometimes weeks ahead of any alarm, avoiding late stage (expensive) fixes. To realize these objectives fully, a proactive management workflow must also be in place. The presentation will discuss new initiatives and workflows developed by plant operations to implement a proactive culture. Real-life case studies from plants using DRA technology will be presented on how this culture change resulted in increased uptime, efficiency and safety of plant operations.

531309 Gaussian Processes for Hybridizing Analytical & Data-Driven Decision-Making

Abstract:

Surrogate models are widely appreciated in process systems engineering [1]. The typical setting focuses on expensive-to-evaluate, possibly uncertain functions. Examples include: modular process simulators [2], integrated gasification combined cycle processes [3], a carbon capture absorber [4], and many other applications, e.g. [5-8]. Resources are typically limited, so effective decision making requires data-efficient learning.

The data science and statistical machine learning communities typically focus on models learned solely from observed data. But chemical engineering applications may also require explicit, parametric models, e.g. modeling known process constraints, operations constraints, and cost objectives [9]. So, work has integrated semi-algebraic functions with those learned from data [10] or developed semi-physical modeling techniques [11, 12].

This presentation surveys the state-of-the-art in hybridizing analytical and data-driven decision making [13]. We consider three probabilistic modeling applications to these hybrid situations:

Design of experiments for model discrimination [14]. We bridge the gap between classical, analytical methods [15] and Monte Carlo-based approaches [16]. Classical methods may have difficulty managing non-analytical model functions and data-driven Monte Carlo approaches come at a high computational cost. We replace the original, parametric models with probabilistic, non-parametric Gaussian process surrogates learned from model evaluations. The surrogates are flexible regression tools that extend classical analytical results to non-analytical models, while providing us with model prediction confidence bounds and avoiding the computational complexity of Monte-Carlo approaches.

Multi-objective optimization [17, 18]. We make novel extensions to Bayesian multi-objective optimization in the case of one analytical objective function and one black-box, i.e. simulation-based, objective function. The resulting method has been applied to a bone neotissue application [19] and a more general test suite.

Scheduling plant operations under uncertainty. For processes with equipment degradation, we use Gaussian processes to approximate large-scale, mixed-integer optimization problems.

We close by offering a broad outlook on applying probabilistic surrogate models to chemical engineering. Statistical machine learning has recently attracted significant interest in process systems engineering [20]. Here we show that state-of-the-art research in Gaussian processes [21, 22] and probabilistic modeling more generally [23] can have a big impact on chemical engineering.

References

[1] Bhosekar A, Ierapetritou M. Advances in surrogate based modeling, feasibility analysis and and optimization: A

review. *Comput Chem Eng*. 108: 250-267, **2018**.

[2] Caballero JA, Grossmann IE. An algorithm for the use of surrogate models in modular flowsheet optimization. *AIChE J*. 54(10):2633-2650, **2008**.

[3] Lang Y, Zitney SE, Biegler LT. Optimization of IGCC processes with reduced order CFD models. *Comput Chem Eng*. 35(9):1705-1717, **2011**.

[4] Cozad A, Sahinidis NV, Miller DC. Learning surrogate models for simulation-based optim-ization. *AIChE J*. 60(6):2211-2227, **2017**.

[5] Boukouvala F, Ierapetritou M. Surrogate-based optimization of expensive flowsheet modeling for continuous pharmaceutical manufacturing. *J Pharm Innov*. 8: 131, **2013**.

[6] Soepyan FB, Cremaschi S, Sarica C, and others. Estimation of percentiles using the Kriging method for uncertainty propagation. *Comput Chem Eng*. 93:143-59, **2016**.

[7] Shokry A, Ardakani MH, Escudero G, Graells M, Espuña A. Dynamic Kriging-based fault detection and diagnosis approach for nonlinear noisy dynamic processes. *Comput Chem Eng*. 106:758-76, **2017**.

[8] Tran AP, Georgakis C. On the estimation of high-dimensional surrogate models of steady-state of plant-wide processes characteristics. *Comput Chem Eng*. DOI 10.1016/j.comp-chemeng.2018.02.014, **2018**.

[9] Boukouvala F, Hasan MF, Floudas CA. Global optimization of general constrained grey-box models: new method and its application to constrained PDEs for pressure swing adsorption. *J Glob Optim*. 67(1-2): 3-42, **2017**.

[10] Boukouvala F, Floudas CA. ARGONAUT: AlgoRithms for Global Optimization of coNstrAined grey-box compUTational problems. *Optim Lett*. 11(5): 895-913, **2017**.

[11] Pearson RK, Pottmann M. Gray-box identification of block-oriented nonlinear models. *J Process Contr*. 10(4):301-315, **2000**.

[12] Cozad A, Sahinidis NV, Miller DC. A combined first-principles and data-driven approach to model building. *Comput Chem Eng*. 73: 116-27, **2015**.

[13] Boukouvala F, Misener R, Floudas CA. Global optimization advances in mixed-integer nonlinear programming, MINLP, and constrained derivative-free optimization, CDFO. *Eur J Oper Res*. 252: 701-727, **2016**.

[14] Olofsson S, Deisenroth MP, Misener R. Design of Experiments for Model Discrimination Hybridising Analytical and Data-Driven Approaches. *arXiv:1802.04170*. **2018**.

[15] Buzzi-Ferraris G, Forzatti P, Emig G, Hofmann H. Sequential experimental design for model discrimination in the case of multiple responses. *Chem Eng Sci*. 39(1): 81-85, **1984**.

[16] Vanlier J, Tiemann CA, Hilbers PAJ, van Riel NAW. Optimal experiment design for model selection in biochemical networks. *BMC Syst Biol*, 8(20), **2014**.

[17] Olofsson S, Mehrian M, Geris L, Calandra R, Deisenroth MP, Misener R. Bayesian Multi-Objective Optimisation of Neotissue Growth in a Perfusion Bioreactor Set-Up. *Computer Aided Chemical Engineering*. 40: 2155-2160, **2017**.

[18] Beykal B, Boukouvala F, Floudas CA, Pistikopoulos EN. Optimal design of energy systems using constrained grey-box multi-objective optimization. *Comput Chem Eng*. DOI 10.1016/j.compchemeng.2018.02.017, **2018**.

[19] Mehrian M, Guyot Y, Papantoniou I, Olofsson S, Sonnaert M, Misener R, Geris L. Maximizing neotissue growth kinetics in a perfusion bioreactor: An *in silico* strategy using model reduction and Bayesian optimization. *Biotechnol Bioeng*. 115(3):617-29, **2018**.

[20] Lee JH, Shin J, Realff MJ. Machine learning: Overview of the recent progresses and implications for the process systems engineering field. *Comput Chem Eng*. DOI 10.1016/j.compchemeng.2017.10.008, **2017**.

[21] Damianou A, Lawrence N. Deep Gaussian processes. In *Artificial Intelligence and Statistics*. 207-215, **2013**.

[22] Deisenroth MP, Ng JW. Distributed Gaussian processes. In *32nd International Conference on International Conference on Machine Learning*. 37: 1481-1490, **2015**.

[23] Ghahramani Z. Bayesian non-parametrics and the probabilistic approach to modelling. *Phil Trans R Soc A*. 371(1984):20110553, **2013**.

AUTHOR INDEX